THE BIG STRAWBERRY BOOK OF
ASTRONOMY

Written by Jeanne Bendick • Illustrated by Sal Murdocca

Strawberry Books, distributed by Larousse & Co., Inc., New York

To Marc and Lincoln

The Big Strawberry Book of Astronomy
Copyright ©1979 by One Strawberry Inc.
All rights reserved.
Printed in Hong Kong

ISBN: Trade 0-88470-058-5
  Library 0-88470-059-3
  Paperback 0-88470-060-7
Library of Congress Catalog Card Number: 78-78321

Contents

THINK BIG	8
Think Big Numbers	8
Think Big Distances	10
Once, People Thought Small	11
PEOPLE HAD SOME FUNNY IDEAS	12
IS SEEING BELIEVING?	15
Moving Through Space	16
YOUR ADDRESS IN SPACE	18
Compare Sizes and Distances	20
Distances Inside the Solar System	20
MAYBE THE SOLAR SYSTEM BEGAN LIKE THIS	21
What Keeps the Planets on the Road?	22
THREE FAMOUS ASTRONOMERS	24
Isaac Newton	24
Tycho Brahe	24
Johannes Kepler	25
THREE HANDY WORDS: DENSITY, MASS, AND WEIGHT	25
THE SUN	26
MERCURY	30
VENUS	31
EARTH	32
What Earth Is Like, from the Inside Out	34
Layers of Air, from Down to Up	35
EARTH'S MOON	36
Phases of the Moon	36
Eclipse!	36
The First Good Look at the Moon	37
MARS	38
THE ASTEROIDS	39
JUPITER	40
SATURN	41
URANUS	42
NEPTUNE	43
PLUTO	43
IDENTIFIED FLYING OBJECTS: COMETS, METEORS, AND METEORITES	44
BEYOND THE SOLAR SYSTEM	46
STARS	46
Stars Are Different from One Another	47
Stars Are Born	48
Stars Die	48
Supernovas	49
Stranger and Stranger	49
Mysteries About Black Holes	50
More Mysteries: Quasars	51
CONSTELLATIONS	52
GALAXIES	54
IS THERE LIFE ON OTHER WORLDS?	56
SOME MYSTERIES ABOUT THE UNIVERSE	58
INDEX	61

Think Big

Think Big Numbers

If you stop to think about what you are seeing when you look up at the sky on a clear night, it might make you feel uncomfortable.

The stars are not just tiny pinpoints of light, but blazing suns. Most of them are much larger than the Earth's sun. And there are billions of them out there in space.

A billion is a thousand million. That's a very big number. But it's a small number in the universe.

There are probably more stars than there are drops of water in all the oceans. A billion stars are less than a drop in the ocean of stars.

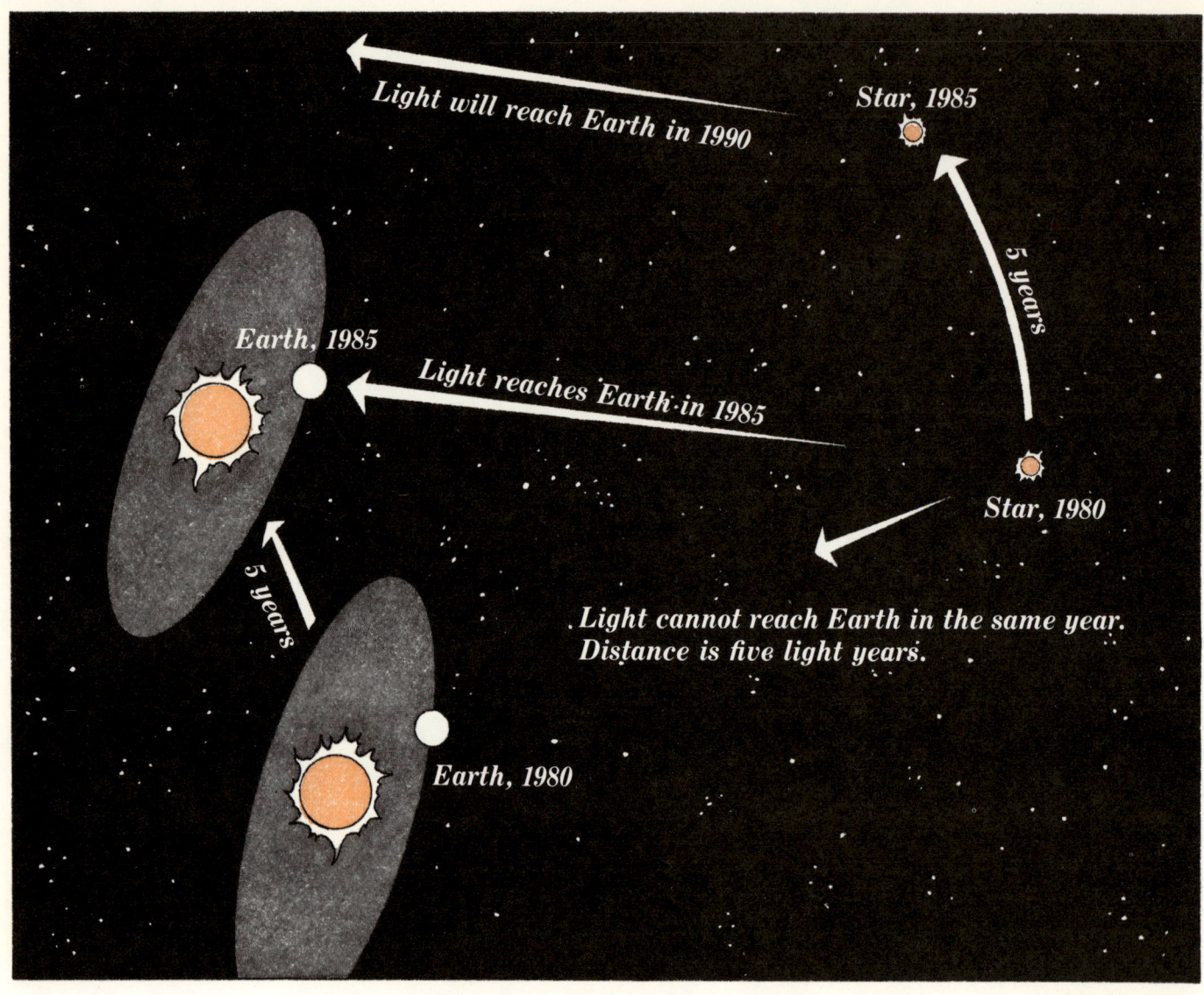

Think Big Distances

When you look out into space at night you are looking across distances bigger than you can imagine. On Earth, a million miles (1.6 million kilometers) is a huge distance. It is about 40 times around the Earth.

But in space, 1.6 million kilometers (a million miles) is a very small distance. Venus, the nearest planet to Earth is about 42 million km (about 26 million miles) away.

Distances in the universe are so enormous that kilometers or miles are not a handy measure. Big distances need big units of measurement.

Scientists use a unit called a light-year. A light-year is a measure of distance and time.

As far as anybody knows, nothing moves faster than light. It moves at a speed of about 300,000 km (186,000 miles) a second.

So, in a year, light moves about 9.5 million million km (about six million million miles). That distance is called a light-year.

Maybe you have never thought about light moving, but it does. For example, it moves across a room, from a light bulb to your eyes. That happens so quickly that it seems to take no time at all.

But when light moves across the immense distances in space, it takes a long, long time, even at 300,000 km (186,000 miles) a second.

Most stars are too far away to see. What you do see is the light each star makes. The light has moved across space from the star to your eyes.

Maybe that took a few years or maybe a thousand years. Maybe the light left that star at the time of the dinosaurs. Or even before.

Once, People Thought Small

People didn't always know that space was so big.

Once they thought that the universe was a small, almost cozy place. Only the Earth was really important. Anyone could see that the Earth was the center of everything.

The Earth looked bigger than the sun, the moon, and the stars. The sun and the moon seemed to move around the Earth. The stars seemed to be fixed like tiny lights in the sky ceiling.

More than 5,000 years ago, careful observers began to keep records of what happened in the sky. That made them able to predict things.

In Egypt, for example, certain stars appeared every year just before the Nile River overflowed. When they saw those stars, the sky watchers said that it was planting time.

In many parts of the world sky watchers could even predict an eclipse. Most people thought they were magicians.

About 2,500 years ago sky watchers began to get new ideas about how the universe worked.

But most people don't like to change their ideas, especially if the new ideas frighten them. When people thought that the Earth was flat and never moved, they didn't want to believe someone who said that it was round and that it was whirling through space.

Would you believe that?

And if you thought that the Earth was the most important thing in the universe, would you believe that it was only a tiny planet, circling a middle-sized star, at the edge of a galaxy that had a hundred billion stars, in a universe where there were billions of galaxies?

Those were uncomfortable ideas. They still are, even though we are sure they are true.

It took many hundreds of years for people to accept some of the ideas we have now about how the universe seems to work.

It still takes a lot of thinking big.

People Had Some Funny Ideas

The ancient Babylonians thought that the Earth was inside a hollow mountain, floating on the sea. Above them, inside the mountain, hung the sun, the moon, and the stars.

The ancient Egyptians thought that the whole Earth was their god Keb, and that the Milky Way was the goddess Nut, bending over the sky.

The ancient Hindus believed that the Earth was a bowl held up by elephants. The elephants stood on a giant turtle, which stood on a snake. When the elephants moved, there were earthquakes.

The very early Greeks thought that the Earth was a flat disk floating on water. Beyond the sky there was fire that showed through little holes in the sky.

Later Greek scientists began to have some other ideas about the Earth, the planets, and the stars.

In about 150 B.C. the great Greek scientist Ptolemy pictured the universe this way:

The Earth was a motionless sphere, suspended in space at the center of the universe. All the other heavenly bodies moved around it in big circles.

Ptolemy also collected all the known facts about the universe and the positions of the stars and planets and published them in a book called the *Almagest*.

In Europe, for almost 1,500 years, people accepted Ptolemy's ideas about the way the universe was arranged. Nobody paid any attention to the *Almagest*.

But Arab astronomers used Ptolemy's book all through the Middle Ages to help them make their own observations and measurements of star positions. (Many stars have Arab names.) Slowly what the Arab astronomers had learned became known in Europe.

In 1533, the Polish astronomer-priest Nicholas Copernicus worked out some ideas of his own.

He decided that the sun was the center of everything, not the Earth. He felt that the Earth was just another planet moving around the sun.

But Copernicus was uneasy about these ideas. They were so different from what everyone believed. They didn't even seem sensible. So they were not published until 1543, when he was dying.

At the end of the 16th century an Italian philosopher, Giordano Bruno, said that the stars were all bodies like the sun, and that they were enormously far away. He said that space and the universe had no beginning or end. Of course nobody believed that. In the year 1600 Bruno was burned at the stake for his ideas.

Nowadays, scientists aren't afraid to suggest new ideas about what the universe is like, even though some of the ideas seem far-out — as far-out as the universe seems to be.

A modern scientist, J. B. S. Haldane, once said that the universe is queerer than anyone can suppose. That is turning out to be true. Nobody really knows what it is like. Even with the most powerful instruments we have only seen a small part of what is out there.

The sky is full of mysteries.

Is Seeing Believing?

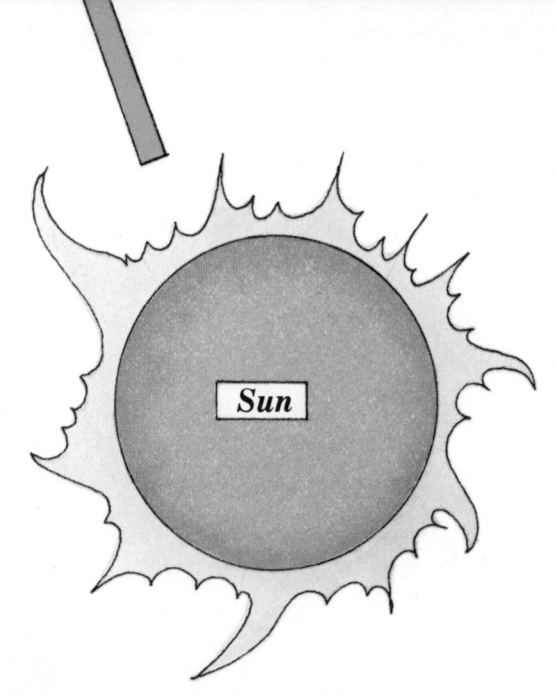

We all depend on our senses to tell us what the world around us is like. But sometimes our senses let us down. You can't always believe what you see or feel.

The Earth looks flat under your feet, but it's not. It's a round ball.

The sun seems to rise in the east every morning, move across the sky, and set in the west. But it doesn't. The sun only *seems* to move because the Earth itself is turning.

The Earth seems still. You can't see or feel it move because everything on it is moving with it.

Moving Through Space

Galaxy

Sun

In fact, everything in the universe is moving.

The Earth spins on its axis. While it is turning, it is also swinging around the sun. All the planets in our solar system orbit the sun and turn as they go.

Some of the planets have moons that move around them. The moons turn as they go, too.

The sun and the other stars make up a great star island in space that is called a galaxy. Our whole galaxy is speeding across the universe. At the same time, all the stars, planets, and moons in the galaxy are turning together, like a huge wheel.

You can't feel any of those motions. You can't feel the Earth spinning. You can't feel it circling the sun, even though it is traveling 96,000 km (60,000 miles) an hour. You certainly can't feel the galaxy moving, though it is going faster than a rocket.

But when you see certain things happening over and over in a regular way, you are really seeing things moving through space.

Night, day, night, day, is the Earth spinning. One spin equals one day. When any part of the Earth is turned toward the sun, it is daytime there.

When a part of the Earth is turned away from the sun, it is night there.

No moon, crescent moon, half moon, full moon, half moon, crescent moon, no moon. That's the moon, circling the Earth every month. Sometimes the whole face is reflecting sunshine Sometimes part of it is in shadow, and that part is dark. Sometimes none of the moon that we see is lit.

Summer, autumn, winter, spring. That is the Earth, going around the sun. It takes a year for the Earth to orbit the sun.

The Earth is tilted on its axis. Where it is tilted toward the sun, as it moves around it, it is summer in that hemisphere. It is winter in the hemisphere that is tilted away.

Days, months, and years show movements through space.

Your Address in Space

    YOUR HOUSE
    YOUR STREET
    YOUR CITY
    YOUR STATE
    YOUR COUNTRY
    YOUR CONTINENT
    EARTH: PLANET 3
    THE SOLAR SYSTEM
    THE MILKY WAY GALAXY
    THE UNIVERSE

The sun and its planets are out here, in one of the pinwheel's arms.

Earth

If we could see the galaxy from the side, it would look like a giant flying saucer of stars.

We live on Earth, the third planet away from a medium-sized star called the sun.

We are on one of nine planets circling that star. All the planets, together with the sun, are called the solar system. ("Solar" means "connected to the sun." Sol was the name of the ancient Greek god of the sun.)

The sun is only one of billions of stars in the galaxy we call the Milky Way. Astronomers think that if we could see our galaxy from the top, it would look like a huge pinwheel made of stars.

The sun and its planets are here.

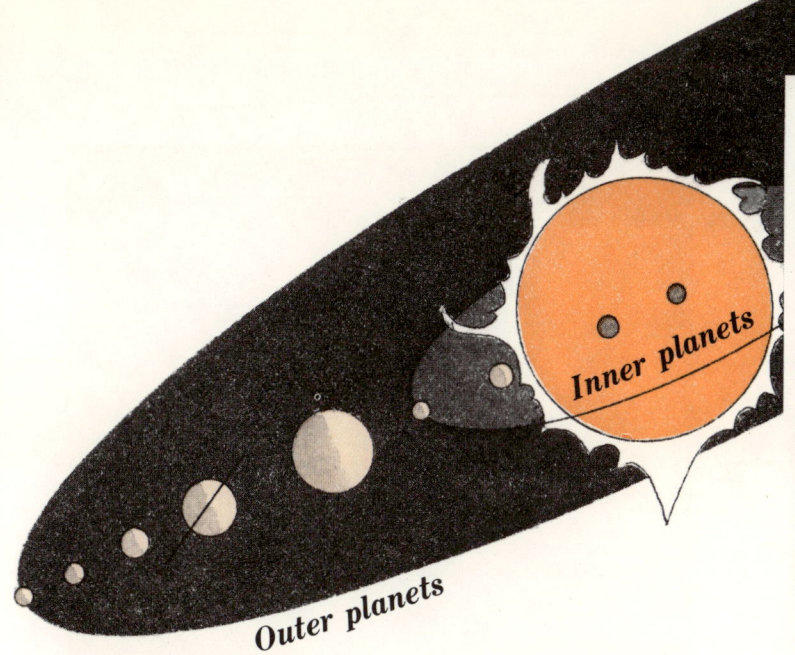

The planets of the solar system swing around the sun, but they stay in their own territories. Each planet has its own path, or orbit, and it stays in that orbit.

The planets closest to the sun are Mercury, then Venus, then Earth, then Mars. These are the inner planets.

Compared to the sun and the outer planets, the inner planets are small. They all have solid, rocky surfaces. Earth and Mars have moons, Mercury and Venus have none.

Beyond the asteroids are the giant planets — Jupiter, Saturn, Uranus, and Neptune — with their many moons.

Beyond the giants is the planet Pluto. It is so far away from the sun that if you were standing on Pluto the sun would look like a very bright star.

Outside the orbit of Mars there is a big space. In this space there are thousands of flying rocks called planetoids or asteroids.

Is Pluto the last planet in our solar system? Most scientists have thought so, but now they are not absolutely sure.

Mercury, Venus, Mars, Jupiter, and Saturn have been known since ancient times, because they can be seen without a telescope. They look like bright, wandering stars. The word "planet" means "wanderer."

Development of a solar system

Compare Sizes and Distances
You can see the huge difference in sizes between the planets. The sun is so much bigger even than Jupiter that there is room here to show only a part of its edge.

Distances Inside the Solar System
The distance from the Earth to the sun is about 149 million km (93 million miles.) That's called one astronomical unit, or 1 AU.

Using that measurement, it's easy to compare the distances of the other planets from the sun.

> Mercury: ⅓ AU
> Venus: ¾ AU
> Mars: 1½ AU
> Jupiter: 5 AU
> Saturn: 9½ AU
> Uranus: 19 AU
> Neptune: 30 AU
> Pluto: 39 AU

Maybe the Solar System Began Like This

Nobody knows for sure how the solar system began. But this is what most scientists think now.

The solar system began as a shapeless cloud of gas floating in space. Maybe that cloud was all that was left of a giant star explosion. Gradually the center of the cloud came together into a roundish shape. More and more gas particles crowded in. The ball of gas got heavier and heavier.

As the particles of gas were jammed together, the center of the sphere kept getting hotter until it began to burn. That was the beginning of our sun. It took about 50 million years for the sun to change from a cloud of gas into a star.

Smaller gas clouds around the sun had begun to come together into other blobs. The blobs began to be shaped like spheres. They got bigger and heavier. They became the planets.

Planets are different from stars. They don't burn like stars. They don't make their own light. The planets in the solar system only reflect the sun's light. The moons of a planet do not make their own light, either. They reflect the sun's light too.

After millions of years, the sun settled down to being a steady star. Its heat and brightness didn't change much.

By about five billion years ago, the sun, the planets, and the planets' moons had the shapes, sizes, and places in space that they have today. They moved around the sun in the same paths that they follow now.

What Keeps the Planets on the Road?
The planets don't zip around the sun every which way. Each planet stays in its own orbit. The sun's gravity holds them there. Gravity is a force that pulls things toward one another. Gravity is a two-way pull. It works on everything in the universe.

Heavy objects have more gravity than light ones. The sun is by far the heaviest object in the solar system. So the sun's gravity pulls on even the farthest planets. Gravity works over huge distances in space.

 The Earth's gravity holds onto everything on and around Earth. It holds onto the air and the water in the oceans. It keeps you from falling off the round Earth, out into space.

 If you drop anything, gravity pulls it down, toward the center of the Earth.

 We can measure the force with which gravity pulls on everything on Earth. It's a familiar measure. We call it weight.

 You can't see gravity. You're so used to it that you can't feel it, either. So how did anyone ever discover it?

Three Famous Astronomers

Isaac Newton

Isaac Newton, who discovered gravity, was born in England in 1642. He began to work out some of his ideas in 1665, when a great plague struck England and his college closed. At home on his mother's farm he had plenty of time to think.

Some stories say that Newton began to think about the idea of gravity when he watched an apple fall to Earth from a tree. He asked an important question about something ordinary: Why does it happen that way? What makes an apple fall down?

One of Newton's big ideas is called the Law of Gravitation. It says that all objects in the universe attract each other, and that the force with which they pull on each other depends on the size and weight of the objects and on how far apart they are.

Newton worked out other famous laws that explain how objects move through space.

Newton didn't suddenly get his ideas about gravity out of a clear blue sky. He had studied the ideas and the mathematical formulas of other scientists who lived before him. When he was famous, he said, "If I have seen further than other men, it is by standing on the shoulders of giants."

Two of those giants were Tycho Brahe and Johannes Kepler.

Tycho Brahe

Tycho Brahe is usually called just Tycho. He was born in Denmark in 1546. In his observatories near Copenhagen, and then near Prague he made very accurate observations and measurements of the motions of the sun, the moon, and the planets, using instruments he had invented himself. (No one had invented the telescope yet.) But he held onto the old idea that the Earth was the center of the universe and the sun moved around it.

Tycho was stubborn, and he got into a lot of fights. He lost the end of his nose in a duel, so he made himself a silver nose to wear.

Johannes Kepler

Johannes Kepler was a German, born in 1571. In 1600 he become Tycho's assitant.

Tycho and Kepler argued about everything, but when Tycho died he left Kepler all his papers. And Kepler used Tycho's observations to prove that Copernicus was right: the planets *did* move around the sun.

Kepler figured out three very important rules about how the planets move: that their paths are not circles, but ellipses; that a planet moves faster when its orbit swings closer to the sun; and that the time it takes for a planet to orbit the sun is related to its distance from the sun.

Kepler was a brilliant mathematician and astronomer. But he also believed in astrology —the idea that the stars predict and control the future. In Kepler's times, most people believed that.

Three Handy Words: Density, Mass and Weight

Density

When the stuff an object is made of is crowded together, the object is dense. A rock is dense. So is a lump of gold and an iron nail. A dense object is heavy for its size, whether it is big or small.

Some things are not dense. A marshmallow is not as dense as a stone. A soap bubble is not as dense as a nail. The bits of matter that make up marshmallows and soap bubbles are farther apart than the bits of matter that make up stones and nails. The density of an object tells you how crowded the bits of matter are in that object.

Mass

Mass is the *amount* of matter in any object. A huge planet like Jupiter is not as dense as the Earth but it has more mass. The size of a thing doesn't tell you about its mass, though. A stone has more mass than a soap bubble, even though the bubble might be bigger. Solid things are usually denser than liquids. Liquids are denser than gases.

Weight

Weight is the measurement of how much gravity is pulling on a thing. Gravity pulls more on heavy things than it does on light things. Dense objects are heavy. They weigh more for their size than less dense objects.

Scientists have ways to measure and compare the density, mass, and weight of planets and stars. This helps them to understand how the solar system is put together and how the universe works.

A red giant is less dense than a white dwarf.

The Sun

Address in the solar system: Right in the middle of everything.

Size: Huge, compared to the planets. Its diameter (the distance through the middle) is about 1,400,000 km (about 865,000 miles).

Our sun is only a medium-sized star, the kind scientists call a yellow dwarf. There are ten billion stars like it in our galaxy, so it's nothing special except to the planets around it.

The sun is a huge, lumpy, flaming ball made mostly of the gases hydrogen and helium. Scientists think that deep inside, where the temperature is millions of degrees, hydrogen is being changed into helium.

When that happens, energy is given off, so the sun is a kind of gigantic nuclear power plant.

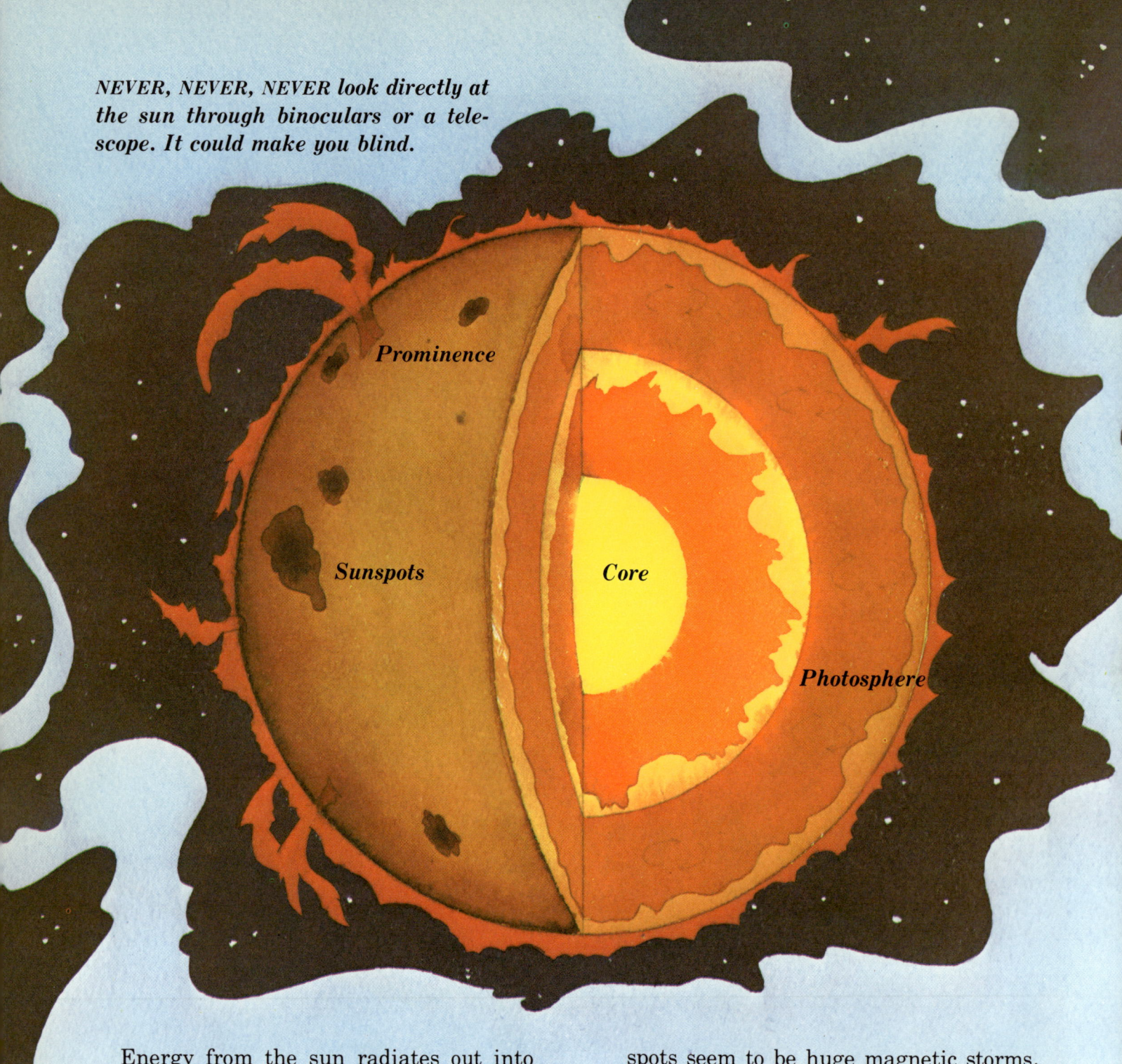

NEVER, NEVER, NEVER *look directly at the sun through binoculars or a telescope. It could make you blind.*

Energy from the sun radiates out into space, mostly as heat and light. But other kinds of energy also come shooting out of the sun — ultraviolet rays, X rays, radio waves, and more. Huge streamers of burning gas flame out in all directions from the sun's atmosphere. An atmosphere is any gas surrounding a star or planet.

The sun rotates, like all bodies in space. But the churning, burning gases that make up the sun do not spin together at the same speed. At the middle, the sun rotates once every 25 days. Near the poles, it takes about 34 days for one rotation.

Some scientists think that the sun's uneven rotation is what causes sunspots. Sunspots seem to be huge magnetic storms.

Sunspots look like big dark spots on the sun. They only look dark because the surface of the sun is so bright. If sunspots were hanging out in space by themselves they would be as dazzlingly bright as spotlights. Big sunspots are much larger than the Earth.

Sunspots may last for weeks or months, and while they are there, great bursts of magnetic energy shoot out from the sun. One burst may release as much energy as people on Earth could use in 100,000 years. Sunspots cause magnetic disturbances all over the solar system. They interfere with shortwave radio transmissions on Earth, and create the northern and southern lights

around the Earth's poles.

Sunspots appear so regularly (about every 11 years) that scientists think they are clocked by some kind of timer deep in the sun.

A great wind of electrified gases called the solar wind blows out of the sun, all the way to the orbit of Pluto, 5,900 million km (3,658 million miles) away.

The sun is about five billion years old. In another five billion years it will start to swell to about 400 times the size it is now. (You'll see why when you read about stars.) It will become a star scientists call a red giant.

By this time, all the inner planets will be burned up. (But you don't have to worry. Five billion years is a very long time from now.)

When the sun has burned up all its fuel, it will begin to cool and shrink into a white dwarf star, no bigger than Earth is now.

Finally, after billions of years of cooling and growing dim, the sun will die and be only a dense, cold, black cinder.

Mysteries About the Sun

What really happens inside the sun?
How hot is the center?
How hot is it around the edges?
Does the sun's atmosphere reach out beyond the solar system?
What causes sunspots?
Why, once in a while, do many years pass without sunspots?

Mercury

Address in the solar system: First planet from the sun.

Distance from the sun: 58 million km (36 million miles).

Mercury is the smallest planet. It is closer in size to our moon than it is to the other planets. Its surface looks a lot like the moon's.

A Mercury year is only 88 Earth days long. That means Mercury's orbit is small. It also rotates slowly. It takes 58.7 Earth days to turn once on its axis. So on Mercury, one day is two-thirds of a year.

Mercury seems to be all rock and metal. It is very dense.

Because it is so small and so near the sun, Mercury is hard to see. But once in a while, if you look low on the horizon just before sunrise or after sunset, you can see it. It looks like a tiny, bright dot.

And because it is so small, Mercury does not have enough gravity to hold an atmosphere. Without an atmosphere, it has nothing to shield it from the sun in the day or to hold the heat at night. So the daytime temperature on Mercury is about 370° C (700° F), and the nights are bitter cold, maybe hundreds of degrees below zero. It wouldn't be a good place to visit. As far as we know, any form of life on Mercury is impossible.

Because it travels so fast, Mercury was named after the speedy messenger of the gods in ancient times.

Mysteries About Mercury

Mercury is smaller than some moons in the solar system. Its landscape looks like our moon's landscape. Is it an escaped moon?

Mercury has hills and ridges cutting across its craters. No other planet has these. What made them?

Venus

Address in the solar system: Second planet from the sun.

Distance from the sun: About 108 million km (about 67 million miles).

Size: Almost the same as Earth—12,320 km (7,700 miles) in diameter. But Venus is not as dense as Earth.

A year on Venus equals about 225 Earth days. Venus rotates so slowly that a Venus day is longer than a Venus year. A Venus day equals 243 Earth days.

Venus and Earth rotate in opposite directions. The only other planet in the solar system that does not rotate in the same direction as Earth is Uranus.

Next to the sun and the moon, Venus is the brightest object we can see in the sky. You can see Venus just before sunrise or just after sunset, but never in the middle of the night. It is always on the sunlit side of Earth.

If there were intelligent life on Venus, the creatures there might never know that the rest of the universe existed, because Venus is entirely surrounded by dense clouds. From the surface, nothing past those clouds would be visible—no sky, no stars, no other planets.

Venus appears so bright and beautiful to us because the clouds around it reflect so much sunlight. Venus was the name of the ancient Roman goddess of beauty.

Once, because of the clouds, people thought that Venus might be a cool, watery planet.

Today we know that Venus is superhot— 485° C (900° F)—and that its atmosphere is made of poison gases.

Mysteries About Venus

Why is Venus so different from Earth, when it is almost Earth's twin in size and place in the solar system? Is Venus at a different stage of development?

In the Venus atmosphere there is a lot of gas called argon that is rare on Earth. Why does Venus have so much argon? Were Venus and Earth formed out of different combinations of gases?

Could the atmosphere of Venus ever change into an Earthlike atmosphere? For more than two billion years, Earth's atmosphere was made of poison gases, too. Earth did not have oxygen until the green planets appeared and began to make it.

Earth

Address in the solar system: Third planet from the sun.

Distance from the sun: About 149 million km (93 million miles).

Size: Diameter at the equator is about 13,000 km (8,000 miles).

Earth is the only planet that isn't named after one of the ancient gods. People had no idea that Earth *was* a planet. It was just the ground under their feet—the earth.

Earth travels around the sun in 365¼ days. That is a year. It turns on its axis once every 23 hours and 56 minutes. That is a day.

Earth was formed at the same time as the other planets, and out of the same stuff. But Earth is special in the solar system.

It is just the right distance from the sun, so that it is warm enough for its water to be liquid. Other planets may have water, but it is frozen into ice.

Its size is important, too. Earth has enough gravity to hold an atmosphere.

Life on Earth depends on liquid water and the kind of atmosphere we have. That is what makes Earth special. As far as we know, it's the only planet in the solar system with life.

The zone around the sun in which a planet can support life is called the ecosphere. Earth is in the middle of the ecosphere.

Venus is at the inner limit of the ecosphere, and Mars, the fourth planet from the sun, is on the outer edge.

Venus is the right size to hold an atmosphere, but its atmosphere is poisonous to life as we know it. Mars is too small—it does not have enough gravity to hold an atmosphere. Earth seems to be the right size and in the right place.

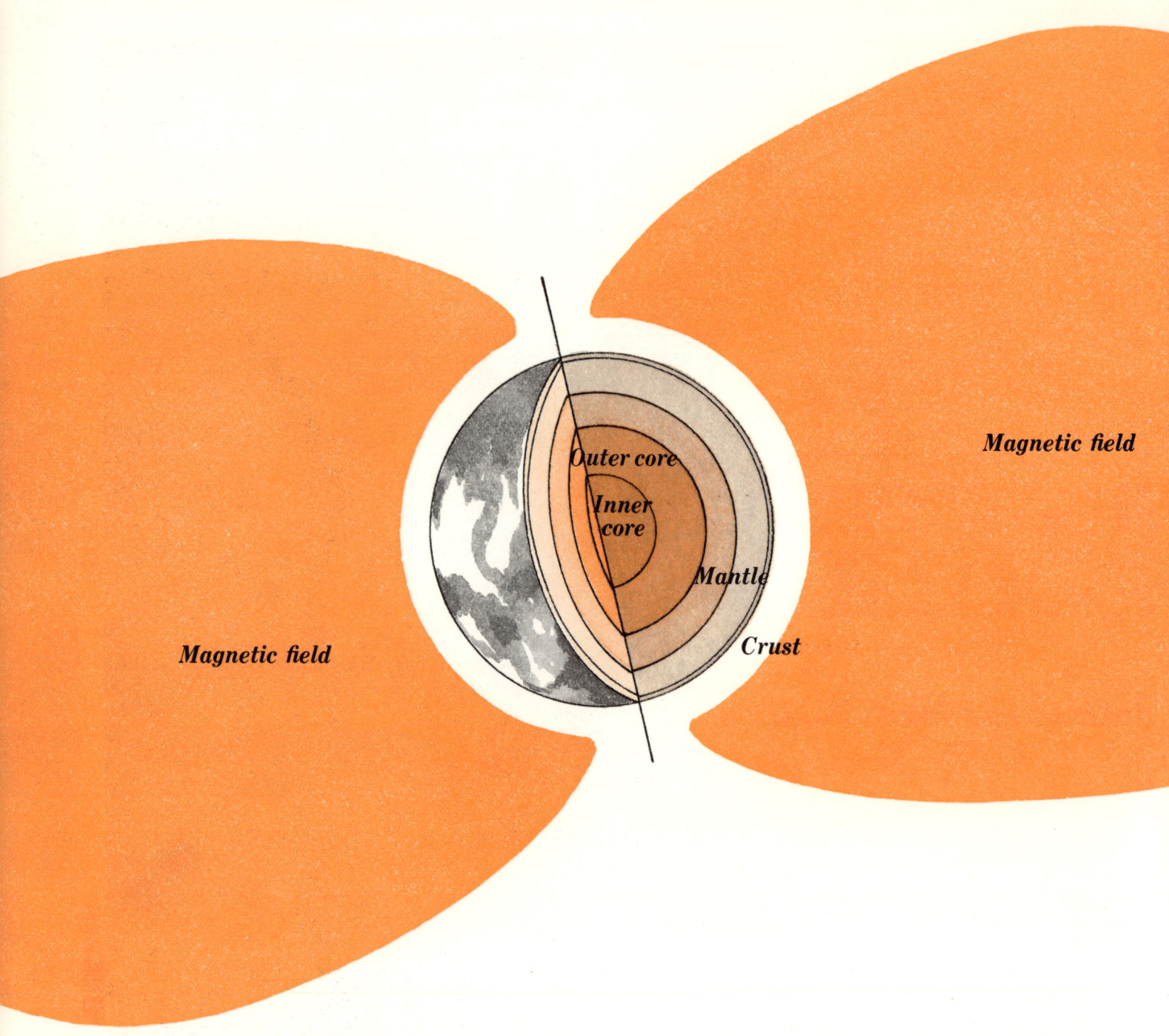

What Earth Is Like, from the Inside Out

Earth is right under our feet, so it's the planet we know the most about. Of course, nobody has been deep down inside. Getting really deep into the Earth is much harder to do than going out into space. Most of what we know about the inside of the Earth we have learned from the instruments that track earthquakes.

The center, or core, of the Earth is flaming liquid rock and iron. That iron core makes the Earth a giant magnet, with north and south poles and a magnetic field like that of an ordinary bar magnet.

Outside the core of the Earth is the mantle, which makes up about 80 percent of Earth's sphere. The mantle is another kind of hot rock. It's not liquid, like the core, but it's not solid either. The mantle is something like very thick bubble gum.

The crust is the top layer of the Earth. The continents are part of the crust. So is the floor of the ocean.

Wrapped around Earth like a blanket is the atmosphere. It protects Earth from the cold of space and the heat and radiations of the sun. The atmosphere keeps us alive. We call it air.

Layers of Air, from Down to Up
The lowest layer of the atmosphere is called the troposphere. It is about three miles thick. Most of the oxygen on planet Earth is in the troposphere, because that's where the plants are. Plants give off oxygen. Animals need oxygen to live.

Mountain climbers on the highest mountains have a hard time breathing, unless they carry oxygen. High-flying airplanes and spaceships have to carry their own oxygen too.

About 13 km (eight miles) up begins the layer called the stratosphere. At the top of this layer there's a kind of oxygen called ozone that screens out a lot of the sun's harmful rays.

About 80 km (50 miles) out from Earth is the layer called the ionosphere. It is magnetic and electrical. At the bottom of the ionosphere the sky is very dark blue. Farther away from Earth it is black, even in the daytime. There is not enough air there to reflect sunlight.

About 400 km (250 miles) out, the exosphere begins. The last trace of the atmosphere is here, going out to about 1200 km (more than 700 miles) from Earth. Then space begins.

The closest thing to us in space is our moon. The moon is about 384,000 km (240,000 miles) from Earth.

Mysteries About Earth
Is Earth the only planet in our galaxy with intelligent life?

What is Earth *really* like, under its crust? No person or instrument has gone down into the mantle.

Scientists think that throughout its history, the Earth's magnetic field has reversed a number of times. What makes that happen?

Earth's Moon

Our moon is not the biggest in the solar system, but except for Pluto's moon it's the biggest compared to its planet.

The moon is about one quarter as large as the Earth. Its diameter is 3,456 km (2,160 miles). The moon is too small to hold an atmosphere. It has no air and no water.

The moon orbits the Earth once every 27.3 days. It also rotates once every 27.3 days. That means that we always see the same side of the moon from Earth. The other side looks pretty much the same.

Once astronomers thought that the moon might have been a part of Earth that broke away. Now they think that Earth and the moon formed at the same time and that Earth's gravity captured the moon.

Of course, the moon has gravity, too, and that makes the tides on Earth. As the moon passes overhead, it pulls on the water in the ocean below. At the place where the moon is pulling, the ocean bulges up. We call that bulge high tide.

When the tide is high in some places, it is low in others, because the high tide water has to come from somewhere.

When the Earth, the moon, and the sun are in a straight line, the combined gravity of the sun and the moon makes the highest tides and the lowest. Those big tides are called spring tides, no matter what time of the year they happen.

When the Earth, the moon, and the sun are at right angles, like this,

the tides are lower than usual.

Because the Earth reflects sunlight too, it has phases like the moon's.

Phases of the Moon

The moon makes no light of its own. It shines because it reflects sunlight. As the moon orbits the Earth, sometimes its whole face is lit by the sun. Sometimes its sunlit side is turned away from us. That's why the moon seems to be a different shape at different times of the month.

Eclipse!

Once in a while the moon, the Earth, and the sun are in an absolutely straight line. Then there is an eclipse.

An eclipse of the moon happens when the moon passes through the Earth's shadow. The sun does not shine on the moon because, for a little while, the Earth is in the way. When the Earth's shadow moves on, the moon shines again.

When the moon is directly between the Earth and the sun, and at exactly the right distance, there is a total eclipse of the sun. From the Earth, the moon appears to blot out the much bigger sun because the moon is closer to us.

Because they kept careful records and were good mathematicians, early astronomers were able to predict eclipses. Since they never said how they did that, people in ancient times thought that astronomers were magicians who could make the sun or the moon disappear and come back.

The First Good Look at the Moon

Galileo was a famous Italian scientist. He made many discoveries about how things move and how they fall.

In 1610 Galileo made himself a telescope to study the sky.

Nobody had ever seen what he saw. He saw that the moon had craters and mountains.

He saw that the planets were not stars. He saw four of Jupiter's moons. He saw that Venus had phases like the Moon. He saw spots on the sun.

Because of what he saw, he challenged some of the old ideas, especially the idea that the Earth was the center of the universe and did not move.

Religious leaders were upset by Galileo's discoveries. After many warnings, Galileo was tried for heresy in 1633 and sentenced to be a prisoner in his house for the rest of his life. But he studied and wrote, even though he went blind from looking at the sun through his telescope. Galileo died in 1642.

Mysteries About the Moon

Is the center hot?

What causes moonquakes?

What made the huge craters on the moon?

Is the moon slowly escaping Earth's gravity?

Mars

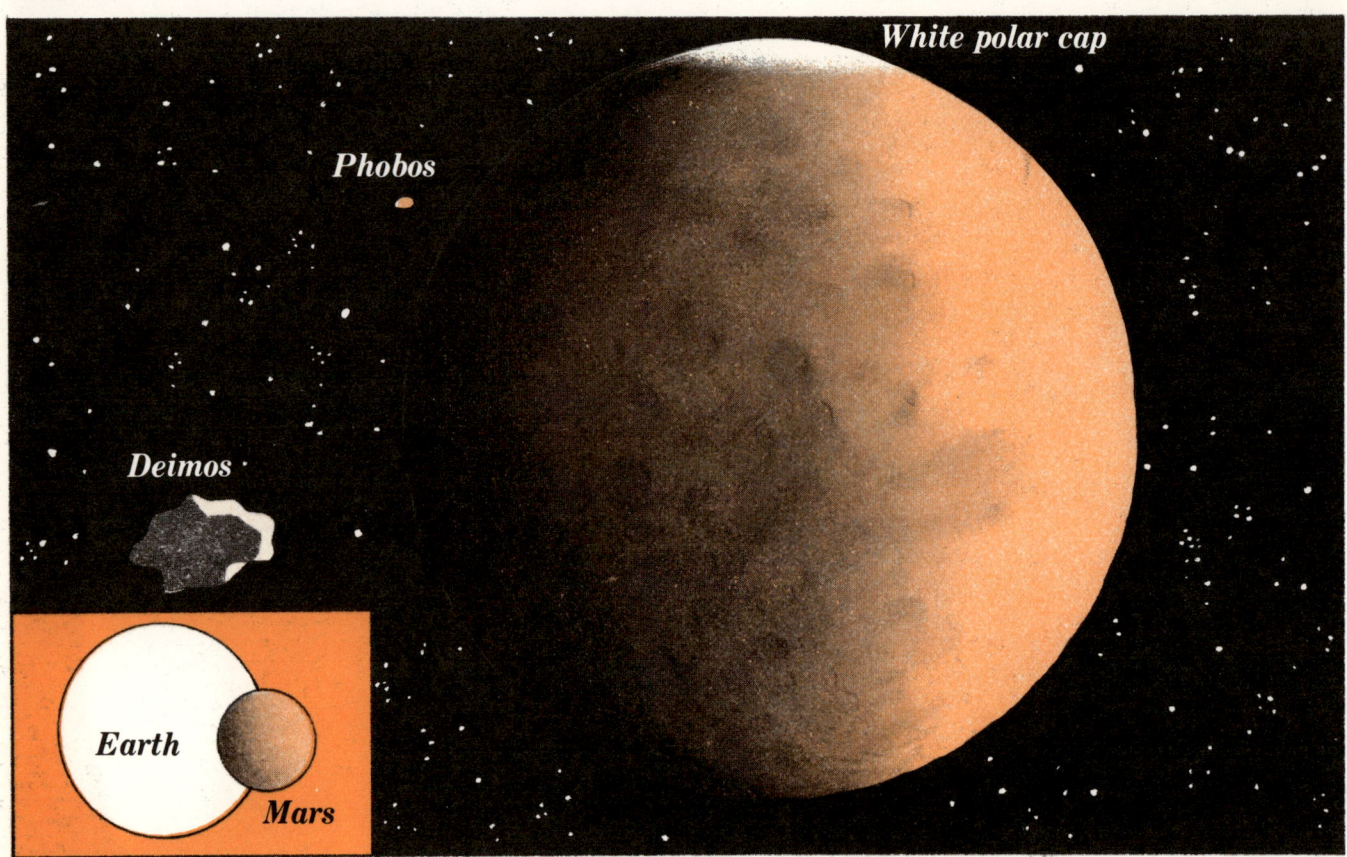

Address in the solar system: Fourth planet from the sun.

Distance from the sun: 228 million km (141 million miles).

Size: Mars' diameter is a little more than half that of Earth, 6,790 km (about 4,220 miles).

A year on Mars is almost twice as long as an Earth year—687 Earth days. But a day is almost the same—24 hours and 37 minutes.

Because Mars is small and light, it does not have enough gravity to hold much of an atmosphere.

There isn't any liquid water on Mars. But there seems to be some water frozen into ice at the poles of the planet.

Not very long ago, many people thought that there might be intelligent life on Mars. There seemed to be a system of canals. Could Martian creatures be drawing water from the poles to other parts of the planet?

Now we know that there are no canals, though it seems as if water did once flow on Mars. There are riverbeds with shapes like islands in them.

There are craters and volcanoes on Mars, too. One volcano, Olympus Mons, is larger than any on Earth.

Most of Mars is a red desert. From Earth, the whole planet looks red. Maybe that's why it was named after Mars, the bloody Roman god of war.

Mars has two very small moons, Phobos and Deimos, which were discovered in 1877. Phobos is very close to the planet—only 9,350 km (5,800 miles) about as far as from New York to Alaska — and it travels very fast, circling the planet more than three times every Martian day.

Deimos, the second moon, is so small and far away from Mars that it would look like a faint star from the surface of the planet.

Both moons seem to be just big, bumpy rocks. Some astronomers think they are captured asteroids.

Mysteries About Mars

Is there even the simplest form of life somewhere on Mars?

What happened to the water on the planet?

Is Phobos getting closer to Mars? Will it crash into Mars someday?

The Asteroids

Between Mars and Jupiter there is a big space of about 560 million km (350 million miles). Astronomers had noticed the space for a long time. In 1772 Johann Bode, a German astronomer, worked out a plan showing the distances of the planets from the sun. According to Bode's Law, there should have been a planet in the space between Mars and Jupiter.

Eighteenth-century astronomers began to look for a missing planet there. They didn't find a full-sized planet, but they did discover some tiny planets that are usually called asteroids. "Asteroid" means "little star."

Most of the asteroids are smaller than grains of sand. Some are pebble-sized. The largest asteroid, Ceres, is much smaller than Earth's moon.

Some people think the asteroids might have been a planet that broke up. Most astronomers think that they were always bits and pieces. There may be 50,000 asteroids or more. If all those flying rocks and pebbles were put together, they would still be smaller than our moon.

Mysteries About the Asteroids

Were they once a planet? If they were, what made the planet break apart?

Some asteroids swing closer to the sun than Mercury. Some swing out past Saturn. Why don't all the asteroids stay in their own zone?

Jupiter

Address in the solar system: Fifth planet from the sun, on the far side of the asteroid belt.

Distance from the sun: About 778 million km (484 million miles).

Size: Enormous—1,300 times bigger than Earth. Its diameter is 143,000 km (89,000 miles) at its equator.

Jupiter is larger than all the other planets combined. It is the giant of the solar system, and is named after the king of the Roman gods.

Even though it is huge, Jupiter is a lightweight. It is much less dense than the Earth. It seems to be made mostly of the gas hydrogen.

A Jupiter year is almost 12 Earth years long, but a Jupiter day is short—nine hours and 51 minutes.

Jupiter looks striped. The stripes are bands of clouds, moving at different speeds. Spots come and go on the stripes, except for one, the Great Red Spot, which is larger than the Earth.

Astronomers have been watching the Great Red Spot for hundreds of years. Sometimes it disappears, but it always comes back. For the last hundred years it has been bright red. A few scientists think it may be a sort of island floating in Jupiter's atmosphere. Most scientists think it is a huge, whirling storm.

Jupiter has 13 moons, and maybe more. Galileo discoverd four with one of his first telescopes. One moon, Ganymede, is bigger than the planet Mercury. Two others, Io and Callisto, are larger than Earth's moon.

Jupiter and its moons almost make up a small solar system by themselves, although Jupiter is not quite hot enough to burn like the sun. Jupiter even has an asteroid belt making a ring around it.

Jupiter has a strong magnetic field. It also sends out strong radio waves.

Mysteries About Jupiter

What is the Great Red Spot?

Why is Jupiter so magnetic?

What makes the radio waves Jupiter sends out?

Why are Jupiter's moons so different from each other? And why do nine of them orbit in one direction and four in the other?

Saturn

Address in the solar system: Sixth planet out from the sun.

Distance from the sun: 1,427 million km (887 million miles).

Size: Big—the second largest planet. Its diameter at the equator is 120,000 km (75,000 miles). Its area is about 80 times greater than Earth's.

Saturn is very cold. Its temperature is about −170°C (−240°F)

Saturn was the farthest planet known in ancient times. It was the last one that could be seen with eyes alone. It was named after the Roman god of agriculture, Saturn.

It takes Saturn 29.46 Earth years to orbit the sun. A Saturn day is about 11 hours long.

Like Jupiter, Saturn is a giant gas ball. It is so light that it would float on water.

The special thing about Saturn used to be its rings. Now we know that at least two other planets, Jupiter and Uranus have rings too.

Saturn's rings are broad enough to see with a small telescope. There are two wide, bright rings and a fainter, dark one closer to the planet.

Saturn has ten moons. The largest, Titan, is much larger than our moon. As far as we know, it is the only moon in the solar system that has an atmosphere. A few of Saturn's other moons seem similar to ours, but smaller. The rest seem to be captured asteroids or big balls of ice.

Mysteries About Saturn

What made Saturn's rings?
Could there be some kind of life on Titan?

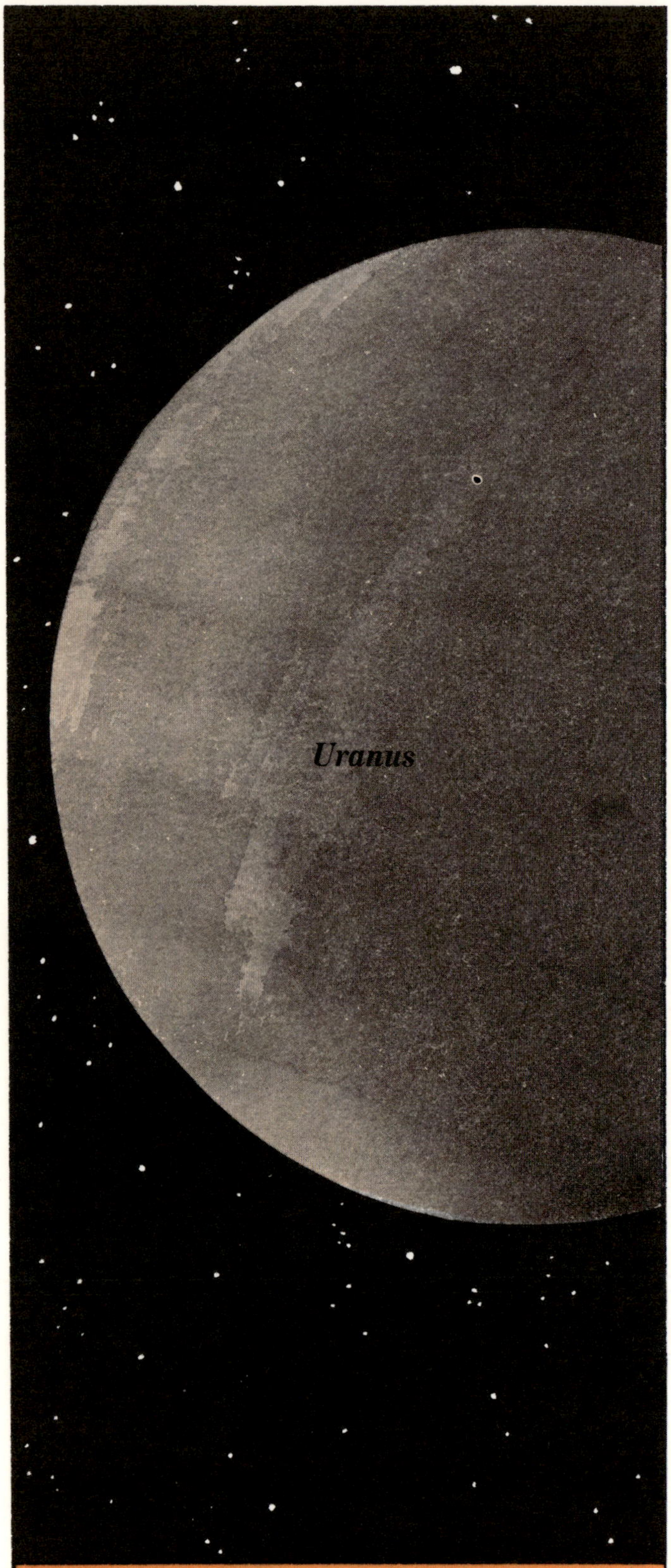

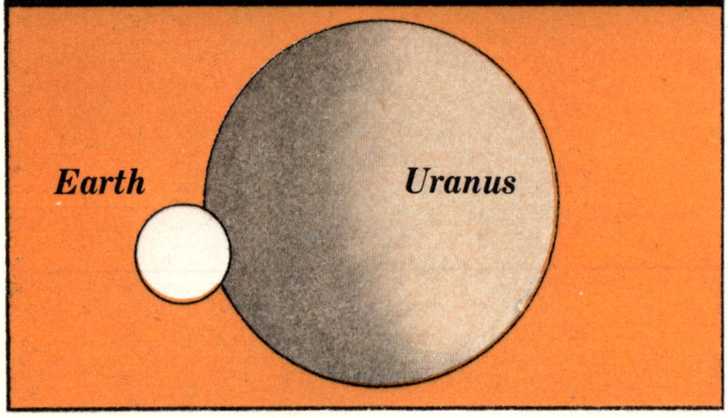

Uranus

Address in the solar system: Seventh planet from the sun.

Distance from the sun: 2,870 billion km (1.78 billion miles).

Size: Compared to Earth, Uranus is a giant. Its diameter is 51,800 km (32,200 miles). But that is only a little more than a third of Jupiter's diameter.

Uranus was the first planet to be discovered with a telescope. It was accidentally discovered by the astronomer William Herschel in 1781.

Uranus was named after the Greek god who gave heat, light and rain to the Earth.

It takes 84 Earth years for Uranus to orbit the sun, which looks like a brilliant star, smaller than our moon but much brighter.

Uranus has a very thick atmosphere. Near the top of the atmosphere it is very cold — about −180°C (−350°F).

Uranus has five moons, all smaller than our moon. Like Saturn, Uranus has rings, but they are very, very thin.

Mysteries About Uranus

Uranus has a very odd tilt. Its axis lies almost flat to its orbit, so it seems to be lying on its side as it rotates.

Why does Uranus speed up, slow down, and wander in its orbit?

How long is a day on Uranus?

Neptune

Address in the solar system: Eighth planet from the sun.

Distance from the sun: 4.5 billion km (2.8 billion miles).

Size: Neptune is similar to Uranus. It is 44,320 km (27,000 miles) in diameter.

Neptune was named after the Roman god of the sea, who was Jupiter's brother.

Astronomers guessed that a planet was there years before Neptune was discovered. They had observed that Uranus had a peculiar orbit, and they thought that the gravity of some unknown planet might be pulling on it.

In 1846 Neptune was located just where it was predicted to be.

A year on Neptune equals almost 165 Earth years. A day seems to be about 15 hours long.

Neptune has two moons. One moon, Triton, is larger than our moon.

Neptune is so far away that astronomers can't see any details of its surface, except that it looks blue. From space the Earth looks blue, too.

Mysteries About Neptune

Why does it look blue?

Why does Neptune have a peculiar orbit, like Uranus?

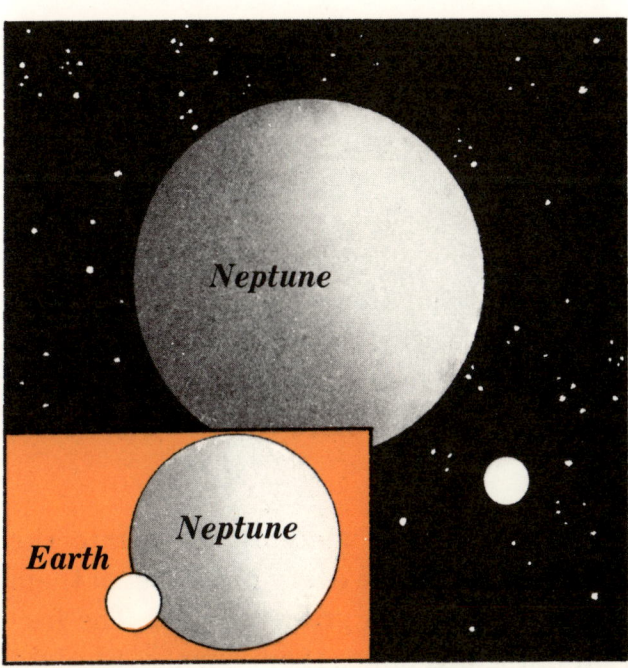

Pluto

Address in the solar system: Ninth planet from the sun.

Distance from the sun: 5.9 billion km (3.67 million miles).

Size: Small. Pluto seems to be about 5,760 km (3,600 miles) in diameter.

As far as we know, Pluto is the outermost planet in the solar system. It was named after the Greek god of the underworld.

Pluto was found in almost the same way that Neptune was. Astronomers wondered what seemed to be affecting the orbits of Uranus and Neptune. After more than 15 years of searching for it, astronomers located Pluto in 1930.

Many astronomers believe that Pluto was not originally a planet, but was a hunk of Neptune, or maybe an escaped moon.

A Pluto year is almost 248 Earth years long. A Pluto day is 6.4 Earth days long.

Astronomers discovered in 1978 that Pluto has a moon that is almost half its own size. This moon is only a little more than 19,000 km (about 12,000 miles) from the planet.

Mysteries About Pluto

Is Pluto really a planet, or just an escaped piece of Neptune?

Since Pluto is so small, astronomers wonder how it could be responsible for the odd orbits of Uranus and Neptune, which are so much larger. Could there be another planet past Pluto?

*Identified Flying Objects:
Comets, Meteors, and Meteorites*

Comets

People used to think that comets were warnings of terrible things to come — plagues, wars, or the death of important people.

There seem to be millions of comets in and beyond the solar system, but only a few of them are big and bright enough to see without a telescope.

The most solid part of a comet is made of tiny, icy particles along with some bits of metal. The tails are mostly gas or dust. Gas tails stream out straight. Dust tails curve.

Some comets have tails millions of kilometers long. No matter which way the comet is going, its tail always points away from the sun.

Comets look bright because they reflect sunlight. Sometimes the gas in the tail glows the way a fluorescent light glows.

Comets travel in long ellipses around the sun. Usually astronomers can figure out their orbits and predict when they are coming. Some may come around twice in a person's lifetime. Some may take thousands of years to orbit the sun. Some may swing out so far they escape the sun's gravity and just vanish into space.

Mysteries About Comets

Where and why do comets begin?
Are they from outside the solar system?
Are they older than the solar system?

Meteors

Many millions of meteors come into the Earth's atmosphere every day. Most of them are no bigger than grains of sand. The friction in the air makes them burn, so at night

they look like streaks of light. Some people call them shooting stars. Meteors seem to be bits of stuff from passing comets, or from comets that have broken up.

There are meteor showers from certain parts of the sky at regular times of the year. Meteor timetables are dependable, so if you watch in the right places at the right times, you might see dozens of shooting stars in an hour.

Mysteries About Meteors

Sometimes a predicted meteor shower never comes. Why?

Meteorites

Meteorites are different from meteors. They are much much larger, and are made of stone or iron.

Meteorites are big enough to make it through the atmosphere and land on Earth. A few hundred fall on Earth every year, but since seven-tenths of the Earth's surface is covered with water, not many fall on land. That's lucky, since a meteorite can weigh many tons. The last really big meteorite to fall on Earth landed in Mexico in 1969. It was brighter than the full moon and was seen over an area of a million square miles. It made sonic booms as it moved through the air and broke into thousands of pieces.

Mysteries About Meteorites

Where do meteorites come from?

Tektites are small, glasslike objects found in Australia, Texas, and a few other places. Are they meteorites? Planetoids? Comets?

Why are they like glass?

Beyond the Solar System

Stars

Just looking at the sky on a clear night, you can see about 2,000 stars. But you don't always see the same stars. Over a year, you might see 6,000 different ones. Some parts of the sky seem jam-packed with stars that look close together, but stars are really very far apart. Alpha Centuri, the star nearest our sun, is 42 million million km (26 million million miles) away — 4.4 light-years.

If the sun were the size of a dot, that nearest star would be a dot ten miles away.

Sirius, the star that looks brightest to us here on Earth, would be another bigger dot twenty miles away. Some stars are millions of light-years away.

Stars shine with the energy they make inside themselves. We see the energy that stars give off as starlight. We call it sunlight when it comes from our own star.

Some stars are hundreds of thousands of times brighter than our sun. The measurement of a star's brightness is called its magnitude.

Magnitude depends on two measurements — how bright the star really is, and how far from us it is. The lower a star's magnitude number is, the brighter the star looks to us. A sixth-magnitude star is very dim. A first-magnitude star is very bright. The sun is so close to us that its magnitude is -27.

White giants (blue)

Red giants

Double star

Stars Are Different from One Another

Stars are masses of gases. If the particles of gas are close together, the stars are dense and heavy. If the particles of gas are far apart, the stars are very light.

About a third of all stars are doubles—two star neighbors revolving around each other.

Some stars are variable. They grow brighter and dimmer, brighter and dimmer over definite periods. They are like lighthouses in space, each one flashing its own special signal. Astronomers use the signals of some variable stars to measure distances in space.

Other stars, called novas, blaze up suddenly and unexpectedly, becoming thousands of times brighter than usual for a few days. Then they settle down again, though they may take years to lose all of their sudden brilliance. Novas may flare up more than once, but they don't do this on any regular schedule.

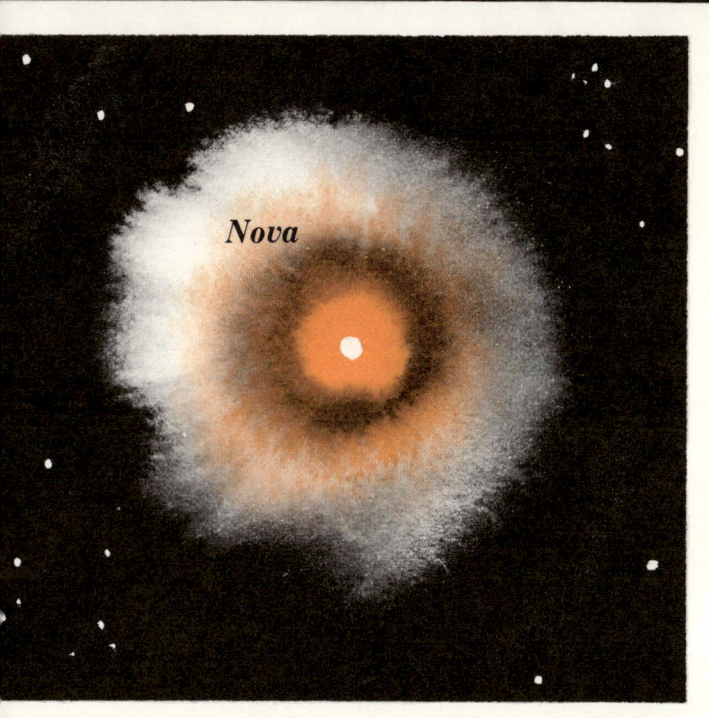
Nova

Some giant red stars are much lighter than air. Some white dwarf stars are so dense that a cup of their starstuff might weigh ten tons. The stars called neutron stars are billions of times heavier than that.

Stars are very different in size. Our sun is an average-sized star. The white dwarfs are no bigger than planets, while the red giants are enormous.

Our sun, compared to the red giant star Antares, is like comparing a pea to a watermelon. Still, there are stars larger than Antares.

White dwarf

Red giant

Stars Are Born
Scientists think that stars are born over millions of years from clouds of dust and gases that come together until they are dense enough and hot enough to start burning.

The youngest stars are big and pale red-orange. As they get older they get smaller and hotter and they change color. Some are yellow, like our sun. White stars are hotter. Blue stars are the hottest. Star colors tell astronomers about the temperature of a star.

Stars in their steady "middle years" are called main-sequence stars.

Stars Die
Stars change over billions of years.

Stars burn from the center. As a star's fuel begins to run low at the center, the outer layers of the star cool and expand. The star swells up and becomes a red giant.

When its fuel is almost gone, a red giant star slowly collapses into a white dwarf. Finally it will be a dead black dwarf, with no heat and no light.

Supernovas

That is what happens to some stars. But other, much bigger stars change faster and in a different way. Instead of changing to a white dwarf, a very big star might explode in a brilliant flash called a supernova. A supernova sends out as much energy in a few seconds as the sun does in a million years or more.

When the explosion is over, the supernova has become an immense cloud of gas around the star that has collapsed until it is smaller and denser than a white dwarf. A star like that is called a neutron star.

Some neutron stars are very magnetic. They send out radio signals in short bursts. They are called pulsars, or radio stars.

Stranger and Stranger

A neutron star is much denser than a white dwarf. Its gravity is immense.

Some neutron stars keep condensing even more. Their gravity becomes so great that nothing — not even light — can escape from them. When this happens, the star becomes a black hole.

Mysteries About Black Holes

Astronomers don't know much about black holes. They have a lot of questions, but, so far, no answers.

Is the pressure in a black hole so great that it could crush itself to nothing?

Is the gravity around black holes so strong that sooner or later they might suck in everything in the universe?

Do black holes ever give out energy? Could one explode and blast out everything it has pulled in?

Could black holes be tunnels to other universes?

Is our whole universe the inside of a black hole, full of stuff that was sucked in from another universe?

More Mysteries: Quasars

Quasars seem to be something like stars and at first, that's what astronomers thought they were. Then, in 1963, astronomers discovered that quasars were immensely distant — farther away than any identified objects in the sky. And the farther away quasars are from our galaxy, the faster they seem to be moving.

Nothing we know of in the universe moves as fast as quasars except light itself.

A single quasar is much brighter than our whole galaxy. A quasar puts out huge amounts of energy.

Nobody is sure what quasars are or even where they are in space. Some astronomers think that quasars may be new galaxies forming. Others think that quasars may mark the beginning and end of the universe, like headlights and taillights on a long train.

Constellations

Many stars look as if they are in groups that stay together as they move across the sky. We call these star groups constellations.

Stars are real. Constellations aren't real. The stars in them aren't together at all. They only look that way because they are in the same part of the sky.

Over thousands of years, people in many parts of the world have made imaginary pictures around the constellations, and made up stories about them.

But even if constellations are imaginary, they are helpful in locating important stars.

Galaxies

A galaxy is a huge collection of stars, a star island in space. Our own galaxy is a flatish disk with spiral arms. It reaches across space for 100,000 light-years. There are about 100 billion stars in our galaxy.

Most people call our galaxy the Milky Way, because the stars in it seem so thick that they look like a great splash of milk. Most astronomers just call our galaxy "the galaxy."

Our whole galaxy is slowly turning through space. Our solar system rotates with the galaxy about once every 225 to 230 million years. That's called a cosmic year.

One cosmic year ago, reptiles were just beginning to be the most important animals on Earth. Dinosaurs were still in the future.

Until this century, most astronomers believed that our galaxy, the Milky Way, was the whole universe.

But more than 100 years ago, one astronomer William Herschel (the man who discovered Uranus) suggested that some of the bright, hazy spots he could see with his telescope might be other galaxies.

In 1925, the American astronomer Edwin Hubble, working with the new 100-inch telescope on Mt. Wilson, measured the distances to some of those hazy patches of stars and proved that they were far outside the Milky Way.

He also discovered that the galaxies were moving apart, and that the farther those galaxies were from us the faster they seemed to be moving away. Now we know that there are billions of galaxies besides our own. Some are 800 billion light-years away.

Galaxies have different shapes. Roundish galaxies seem to be made of old stars.

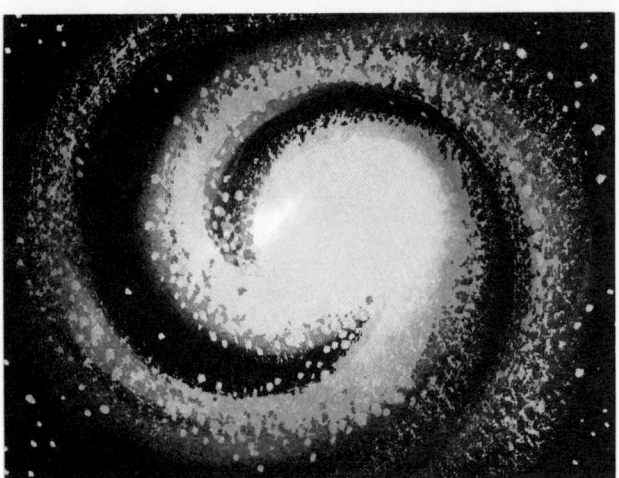

Spiral galaxies seem to be made of the newer stars.

All galaxies send out radio waves.

The radio waves from some galaxies are very powerful. Astronomers think that the waves may have started with tremendous explosions in those galaxies a long time ago. Nobody knows when the explosions occurred, or why.

Horse nebulae

Galaxies are not made just of stars. They also contain clouds of gases called nebulas.

Some nebulas are dark. They look like holes in the sky, but there are probably stars behind them.

Some nebulas are bright. They may reflect the light of nearby bright stars. Or the heat of the stars in the galaxy may make them fluorescent.

Some bright nebulas are all that is left of a supernova that scattered starstuff into space around it, or of a red giant star, puffing off gases as it died.

Astronomers think that sometimes the gases in a nebula may come together to form new stars. Some of those new stars might contain material from old stars that died long ago. Some astronomers think that our solar system might have been formed from what was left of an old supernova explosion.

Is There Life on Other Worlds?

There are billions of galaxies, and each has billions of stars. In our own galaxy, there are ten billion stars like our sun. It is likely that many of those stars have planets, too. Astronomers say that there may be a billion planets in our galaxy that have life on them.

A billion planets with life! Then why haven't we made contact? One reason is the huge distances in space.

The nearest star that *might* have a planet like ours is Tau Ceti, 11 light-years away. Even to say hello to a Tau Ceti planet by radio — which travels the speed of light — would take 11 years.

If the nearest civilization is on a planet that belongs to a farther star, a radio conversation might take hundreds or thousands of years.

Our fastest space probe goes about 108,000 km (67,000 miles) an hour. At that speed, it would take a spaceship about 100,000 Earth years to reach Tau Ceti.

Maybe a civilization much more advanced than ours has found a way to travel as fast as

light, or has even found a shortcut through space. Maybe beings from another planet are on the way to Earth?

Would they look like us? Would life on other worlds be like life on Earth?

There is no reason why it should be. Even on Earth, living things are different from each other. Life forms are adapted in many ways to the many different environments on Earth.

So creatures from desert planets, or water-covered planets, or very cold or very hot planets, or huge planets with tremendous gravity, would be very different from Earth people. We might not even recognize them. They might look like carrots or stones or spiders or like puffs of gas.

If intelligent creatures in weird shapes landed on Earth, how would you be able to accept them?

Do you think there is other life out there, somewhere?

Or do you think, that in all the universe, we are the only ones?

Some Mysteries About the Universe

Ideas about how the universe is made have changed many times in the history of sky watching. Once new ideas were dangerous. Anyone who suggested them was likely to be put in prison, or even killed.

Now scientists consider new ideas even if they seem ridiculous, because scientists know that the universe *is* stranger than anyone can suppose.

Albert Einstein's ideas about space and time, matter and gravity in the universe seemed very queer in 1915. Now almost everyone accepts them as true.

It takes a long time to prove or disprove ideas about something that is happening millions of light-years of space and time away. Maybe those ideas never can be proved. So the universe is still full of mysteries.

The Big Bang made new elements that condensed to form new galaxies.

Albert Einstein

Gravity begins to pull the universe together again.

How did the universe begin?

Right now, most scientists think that about 20 billion years ago there was nothing but a small, dense mass of unimaginably hot, bright stuff, that exploded with a big bang.

Out of that came star clouds, which became stars and planets and galaxies. But what was before the big bang?

What is happening to the universe? We know that now the galaxies are moving away from each other. Will the galaxies rush apart forever, out into unending space?

Or, sooner or later, will gravitation slow the galaxies down, then pull them all back together again.

Then what will happen?

Will there be another big bang, so that the universe starts all over?

Or will everything end up in a black hole?

Some scientists think that none of those things will happen. They think that the universe is steady. They point out that old stars die, but new stars are always being born out of the gases in the galaxies and between them. Those scientists think that the universe has no beginning and no end.

Is there a beginning and end of time?

Is there a beginning and end of space?

Are there other universes?

What do you think?

Don't Stop Here.

In a way, this book is an unfinished book because our exploration of the universe keeps on going. Our ideas about how it is made and what is happening out there keep changing. Ideas have changed more in the last few years than they did in all the years from Ptolemy to Copernicus.

Even our own neighborhood, the solar system, is full of surprises. Nobody knows what we will find as we explore the universe.

The universe belongs to everybody, not just to astronomers. Why don't you keep your own big book of what we are all finding out? New ideas make new mysteries. Maybe you will help to solve them.